Teach Multiplication, Division, and the Time Table All at the Same Time

An Instructional Guide for Learning Basic Math Skills

Andray McCuien

Master of Special Education

With an Institutional Recommendation

Learning Disabilities Endorsement

This guide can be used as a supplement for remedial math, for struggling students, for enrichment, or as resource material.

AuthorHouse™
1663 Liberty Drive
Bloomington, IN 47403
www.authorhouse.com
Phone: 1-800-839-8640

First published by AuthorHouse 4/19/2011

ISBN: 978-1-4567-2770-3 (sc)
ISBN: 978-1-4567-4012-2 (e)

Library of Congress Control Number: 2011902165

Printed in the United States of America

Preface

This book can help students develop basic math skills that serve as the building blocks for all mathematical equations and problem-solving concepts. Once a student has mastered his or her basic math skills, adding new formulas and problem-solving concepts simply becomes a matter of understanding when and what sequence to apply for each order of operation. If students do not learn the basic foundation of mathematic- addition, subtraction, multiplication, division, and the time table chart - they will always struggle to develop a comprehension of and excel in math.

This book is intended as a simple guide to help third-grade students become proficient in their application of basic math skills. Students will learn to use addition, subtraction, multiplication, division and the time table chart simultaneously while they learn the strategy of counting up.

I have nothing against the usage of calculators, but when students are introduced to the calculator at the third-grade level, they become dependent on the use of calculators for assistance with computations. Once students learn to rely on the calculator, it becomes unnecessary for them to master basic math skills. The calculator becomes the problem-solving tool. Could this be part of the reason students are not performing at levels of proficiency required on standardized test? It is important for students to conquer basic math skills prior to being introduced to the calculator.

Once students have learned the strategy of counting up, they can rely on their own knowledge, rather than a gadget, to solve math problems. Students will use counting, addition, subtraction, multiplication, division, and reasoning skills as they apply the strategy. Through repetition, practice, and daily drills, students will develop basic math skills, a sense of self- confidence, and organizational independence.

How Students Learn

Teaching students with learning disabilities has helped me to understand that all children have the capacity to learn if given the necessary tools. Using the strategy of counting up provides students with a visual, auditory, and a trial-and-error approach to learning. The visual approach gives students the picture that they need (via their fingers) to identify missing factors in multiplication and division equations; an auditory approach is provided when students count aloud to themselves, which reinforces what the child is learning a second time; and the trial-and-error approach provides the child with the opportunity to search for the correct response and, through practice, learn to identify the correct response on a progressive and consistent basis. Over a period of time, students will develop their proficient use of these approaches. Students are also provided with an opportunity to use self-correction, checking to ensure that the correct number of fingers has been used to count up to the accurate product or response.

Students are also provided with another reinforcement tool: writing out the answers manually to each group of multiplication factors (1–12) until they can identify correct responses as they solve problems. This provides the child with another visual learning tool. Because a learning tool is, an opportunity toward learning.

It is very important for students to learn how to find the answer, rather than memorize the answer itself. When a strategy is learned, it is neatly tucked away within the files of our brain until it is recalled for use. If students simply memorize information, it is filed within our brain on a temporary basis and will soon be forgotten.

In order for students to learn, information must be repeated often, consistently, daily, and on a routine basis over a period of time. Assistance and immediate feedback must be provided in a positive manner as a motivation tool to help students learn the concept. Once students grasp the concept, they will become more productive and work at a more rapid pace. Never discourage students with impatience or negative comments. Always give students a response that causes them to feel as if they have been successful and have accomplished some task that day.

The strategy of counting up consists of using tools students are already comfortable with in a new, progressive way. Over the years, I have taught various grade levels ranging from kindergarten up to twelfth grade; I have observed that most, if not all, students use their fingers to solve math problems. The practice of counting fingers continues into middle school and even high school for some students. This same technique of counting fingers has been developed into a strategy to help students learn basic math skills. Once students become proficient with basic math skills, the use of their fingers will no longer be necessary.

Reader Notes

The Reader is encouraged to make notes as they read this book, learn the strategies and apply them in the classroom setting, one-on-one tutoring, or during math drills. It is often very helpful to make notations of areas of a book to review again a second time or specific strategies that can be applied to help students learn more efficiently.

Things I want to review:

Things I want to remember:

Reader Notes

Things I want to apply:

Things I want to share:

Chapter 1
Multiplication and the Time Table Chart

At this point, students learn the rules for multiplying by 0 and 1.
Rule for multiplying by 0:
Any number multiplied by 0 is equal to 0.

$0 \times 1 = 0$	$0 \times 4 = 0$	$0 \times 7 = 0$	$0 \times 10 = 0$
$0 \times 2 = 0$	$0 \times 5 = 0$	$0 \times 8 = 0$	$0 \times 11 = 0$
$0 \times 3 = 0$	$0 \times 6 = 0$	$0 \times 9 = 0$	$0 \times 12 = 0$

Rule for multiplying by 1:
Any number multiplied by 1 is equal to that same number.

$1 \times 1 = 1$	$1 \times 4 = 4$	$1 \times 7 = 7$	$1 \times 10 = 10$
$1 \times 2 = 2$	$1 \times 5 = 5$	$1 \times 8 = 8$	$1 \times 11 = 11$
$1 \times 3 = 3$	$1 \times 6 = 6$	$1 \times 9 = 9$	$1 \times 12 = 12$

To be sure students learn these two rules, have them write out the entire time table, multiplying each number from 1 to 12 by 0, showing the product of 0 for each set of two factors; have them repeat the same step to illustrate the second rule, multiplying each of the factors by 1 and showing the product.

These two rules are very simple and are usually easy for students to learn. It is usually not necessary to use the counting-up strategy to help students understand these two principles.

Now it is time to move on to the rest of the time table chart. Do not focus on how much of the time table chart students are able to remember at this point. This exercise is simply to help students become acquainted with the time table chart as a concept. Once students begin to apply the strategy of counting up, they will learn the time table chart in its entirety through repetition and practice.

Students should be given a time table chart to use when multiplication is introduced. This is simply to provide students with a visual aid to follow as the time tables are discussed.

Review each factor of the chart in its entirety. Because the rule of multiplying using the multiple of 1 has already been introduced, it should be reviewed briefly. Move on to the 2s, the 3s, and so forth, in sequence, until the entire time table chart from 1 to 12 has been introduced. Copies of the time table charts should not be used when students are applying the strategy of counting up.

Students tend to rely on the chart to determine the product of multiples rather than applying any type of problem-solving strategy. The charts should not be left in the possession of students, but passed out when the time tables are introduced and then collected at the conclusion of the lesson.

Time Table Chart and Basic Principles of Multiplication

Oral practice drills should be a 5- to 10-minute segment of daily instruction.

Times Table Chart 1–12

1 × 1 = 1	2 × 1 = 2	3 × 1 = 3	4 × 1 = 4	5 × 1 = 5	6 × 1 = 6
1 × 2 = 2	2 × 2 = 4	3 × 2 = 6	4 × 2 = 8	5 × 2 = 10	6 × 2 = 12
1 × 3 = 3	2 × 3 = 6	3 × 3 = 9	4 × 3 = 12	5 × 3 = 15	6 × 3 = 18
1 × 4 = 4	2 × 4 = 8	3 × 4 = 12	4 × 4 = 16	5 × 4 = 20	6 × 4 = 24
1 × 5 = 5	2 × 5 = 10	3 × 5 = 15	4 × 5 = 20	5 × 5 = 25	6 × 5 = 30
1 × 6 = 6	2 × 6 = 12	3 × 6 = 18	4 × 6 = 24	5 × 6 = 30	6 × 6 = 36
1 × 7 = 7	2 × 7 = 14	3 × 7 = 21	4 × 7 = 28	5 × 7 = 35	6 × 7 = 42
1 × 8 = 8	2 × 8 = 16	3 × 8 = 24	4 × 8 = 32	5 × 8 = 40	6 × 8 = 48
1 × 9 = 9	2 × 9 = 18	3 × 9 = 27	4 × 9 = 36	5 × 9 = 45	6 × 9 = 54
1 × 10 = 10	2 × 10 = 20	3 × 10 = 30	4 × 10 = 40	5 × 10 = 50	6 × 10 = 60
1 × 11 = 11	2 × 11 = 22	3 × 11 = 33	4 × 11 = 44	5 × 11 = 55	6 × 11 = 66
1 × 12 = 12	2 × 12 = 24	3 × 12 = 36	4 × 12 = 48	5 × 12 = 60	6 × 12 = 72

7 × 1 = 7	8 × 1 = 8	9 × 1 = 9	10 × 1 = 10	11 × 1 = 11	12 × 1 = 12
7 × 2 = 14	8 × 2 = 16	9 × 2 = 18	10 × 2 – 20	11 × 2 = 22	12 × 2 = 24
7 × 3 = 21	8 × 3 = 24	9 × 3 = 27	10 × 3 = 30	11 × 3 = 33	12 × 3 = 36
7 × 4 = 28	8 × 4 = 32	9 × 4 = 36	10 × 4 = 40	11 × 4 = 44	12 × 4 = 48
7 × 5 = 35	8 × 5 = 40	9 × 5 = 45	10 × 5 = 50	11 × 5 = 55	12 × 5 = 60
7 × 6 = 42	8 × 6 = 48	9 × 6 = 54	10 × 6 = 60	11 × 6 = 66	12 × 6 = 72
7 × 7 = 49	8 × 7 = 56	9 × 7 = 63	10 × 7 = 70	11 × 7 = 77	12 × 7 = 84
7 × 8 = 56	8 × 8 = 64	9 × 8 = 72	10 × 8 = 80	11 × 8 = 88	12 × 8 = 96
7 × 9 = 63	8 × 9 = 72	9 × 9 = 81	10 × 9 = 90	11 × 9 = 99	12 × 9 = 108
7 × 10 = 70	8 × 10 = 80	9 × 10 = 90	10 × 10 = 100	11 × 10 = 110	12 × 10 = 120
7 × 11 = 77	8 × 11 = 88	9 × 11 = 99	10 × 11 = 110	11 × 11 =121	12 × 11 = 132
7 × 12 = 84	8 × 12 = 96	9 × 12 = 108	10 × 12 = 120	11 × 12 = 132	12 × 12 = 144

Practice Page

Times Table Chart 1–12
Fill in the missing product.

1 × 1 =	2 × 1 =	3 × 1 =	4 × 1 =	5 × 1 =	6 × 1 =
1 × 2 =	2 × 2 =	3 × 2 =	4 × 2 =	5 × 2 =	6 × 2 =
1 × 3 =	2 × 3 =	3 × 3 =	4 × 3 =	5 × 3 =	6 × 3 =
1 × 4 =	2 × 4 =	3 × 4 =	4 × 4 =	5 × 4 =	6 × 4 =
1 × 5 =	2 × 5 =	3 × 5 =	4 × 5 =	5 × 5 =	6 × 5 =
1 × 6 =	2 × 6 =	3 × 6 =	4 × 6 =	5 × 6 =	6 × 6 =
1 × 7 =	2 × 7 =	3 × 7 =	4 × 7 =	5 × 7 =	6 × 7 =
1 × 8 =	2 × 8 =	3 × 8 =	4 × 8 =	5 × 8 =	6 × 8 =
1 × 9 =	2 × 9 =	3 × 9 =	4 × 9 =	5 × 9 =	6 × 9 =
1 × 10 =	2 × 10 =	3 × 10 =	4 × 10 =	5 × 10 =	6 × 10 =
1 × 11 =	2 × 11 =	3 × 11 =	4 × 11 =	5 × 11 =	6 × 11 =
1 × 12 =	2 × 12 =	3 × 12 =	4 × 12 =	5 × 12 =	6 × 12 =

7 × 1 =	8 × 1 =	9 × 1 =	10 × 1 =	11 × 1 =	12 × 1 =
7 × 2 =	8 × 2 =	9 × 2 =	10 × 2 =	11 × 2 =	12 × 2 =
7 × 3 =	8 × 3 =	9 × 3 =	10 × 3 =	11 × 3 =	12 × 3 =
7 × 4 =	8 × 4 =	9 × 4 =	10 × 4 =	11 × 4 =	12 × 4 =
7 × 5 =	8 × 5 =	9 × 5 =	10 × 5 =	11 × 5 =	12 × 5 =
7 × 6 =	8 × 6 =	9 × 6 =	10 × 6 =	11 × 6 =	12 × 6 =
7 × 7 =	8 × 7 =	9 × 7 =	10 × 7 =	11 × 7 =	12 × 7 =
7 × 8 =	8 × 8 =	9 × 8 =	10 × 8 =	11 × 8 =	12 × 8 =
7 × 9 =	8 × 9 =	9 × 9 =	10 × 9 =	11 × 9 =	12 × 9 =
7 × 10 =	8 × 10 =	9 × 10 =	10 × 10 =	11 × 10 =	12 × 10 =
7 × 11 =	8 × 11 =	9 × 11 =	10 × 11 =	11 × 11 =	12 × 11 =
7 × 12 =	8 × 12 =	9 × 12 =	10 × 12 =	11 × 12 =	12 × 12 =

Chapter 2
Skip Counting

Now it is time to introduce students to the concept of skip counting by the numbers 2, 3, 5, and 10. Students must learn to add the number with which they are skip counting to each preceding product in sequence, using time table multiples, beginning with 1 and ending with 12. Students will learn to identify only the product of each set of multiples 1–12 multiplied by the table of 2, 3, 5, and 10.

Skip Counting by 2
Time Table of 2

$2 \times 1 = 2$	$2 \times 4 = 8$	$2 \times 7 = 14$	$2 \times 10 = 20$
$2 \times 2 = 4$	$2 \times 5 = 10$	$2 \times 8 = 16$	$2 \times 11 = 22$
$2 \times 3 = 6$	$2 \times 6 = 12$	$2 \times 9 = 18$	$2 \times 12 = 24$

Students will practice repeating the product of each set of multiples from the table of 2 in sequence until it is learned. Students are able to gesture using their fingers as they count aloud.

Practice Sequence
2   4   6   8   10   12   14   16   18   20   22   24

Skip Counting by 3
Time Table of 3

$3 \times 1 = 3$	$3 \times 4 = 12$	$3 \times 7 = 21$	$3 \times 10 = 30$
$3 \times 2 = 6$	$3 \times 5 = 15$	$3 \times 8 = 24$	$3 \times 11 = 33$
$3 \times 3 = 9$	$3 \times 6 = 18$	$3 \times 9 = 27$	$3 \times 12 = 36$

Students will practice repeating the product of each set of multiples from the table of 3 in sequence until it is learned. Students are able to gesture using their fingers as they count aloud.

Practice Sequence

3   6   9   12   15   18   21   24   27   30   33   36

Skip Counting by 5
Time Table of 5

$5 \times 1 = 5$	$5 \times 4 = 20$	$5 \times 7 = 35$	$5 \times 10 = 50$
$5 \times 2 = 10$	$5 \times 5 = 25$	$5 \times 8 = 40$	$5 \times 11 = 55$
$5 \times 3 = 15$	$5 \times 6 = 30$	$5 \times 9 = 45$	$5 \times 12 = 60$

Students will practice repeating the product of each set of multiples from the table of 5 in sequence until it is learned. Students are able to gesture using their fingers as they count aloud. While students may stop at 60, the illustrations continue up to 100.

Practice Sequence

5   10   15   20   25   30   35   40   45   50   55   60   65   70

75   80   85   90   95   100

Skip Counting by 10
Time Table of 10

$10 \times 1 = 10$	$10 \times 4 = 40$	$10 \times 7 = 70$	$10 \times 10 = 100$
$10 \times 2 = 20$	$10 \times 5 = 50$	$10 \times 8 = 80$	$10 \times 11 = 110$
$10 \times 3 = 30$	$10 \times 6 = 60$	$10 \times 9 = 90$	$10 \times 12 = 120$

Students will practice repeating the product of each set of multiples from the table of 10 in sequence until it is learned. Students are able to gesture using their fingers as they count aloud.

Practice Sequence

10   20   30   40   50   60   70   80   90   100   110   120

Chapter 3
Introducing Long Division

After students become familiar with multiplication, division can be introduced. I prefer to teach students long division using pencil and paper. It is very important to introduce students to key terms that are associated with division, such as *inverse operations* (operations that undo each other). The rules for dividing by 0 and 1 must also be introduced at this time.

Once students have become familiar with the terms associated with division, allow students to begin manual practice by assisting them as they solve several long-division problems using scratch paper and a pencil. Note the example below.

Long Division

$$6 \overline{\smash{\big)}\,42}$$

Step 1: Help students determine the target number for the equation. The target number in this equation is 42.

Step 2: Students determine which multiplication table will be used to solve the equation. In this equation, the table of 6 will be used, because it must be determined how many times 6 will divide into 42.

Step 3: Students write out the first multiple of the table of 6, which will be 1. (This step of writing out the table of 6 will continue until the missing multiple is identified that will produce a product of 42 when multiplied by 6.)

$6 \times 1 = 6$

Students apply the strategy of counting up using their fingers to help them determine the next product, of 6 times 2.

$6 \times 2 = ?$

Students begin with the number 6 and count up 6 spaces, raising one finger as each number is counted aloud until 6 spaces have been counted.
6,

7   8   9

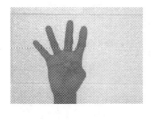

| 10 | 11 | 12 |

Students have counted up 6 spaces from the product of 6 x1 = 6, so they know that $6 \times 2 = 12$. Students manually list the new product.

$6 \times 1 = 6$

$6 \times 2 = 12$

Now students begin the process again to determine the product of the next factor, which is 3. To determine the new product, students will count 6 spaces up from the product of 12. Students will raise one finger as they count up each space.

$6 \times 3 = ?$

12,

| 13 | 14 | 15 |

| 16 | 17 | 18 |

Students have counted up 6 spaces from the product of $6 \times 2 = 12$, so they know that $6 \times 3 = 18$. Students manually list the new product.

$6 \times 1 = 6$

$6 \times 2 = 12$

$6 \times 3 = 18$

Now students begin the process again to determine the product of the next factor, which is 4. To determine the new product, students will count up 6 spaces from the product of 18. Students will raise one finger as they count up each space.

$6 \times 4 =$

18,

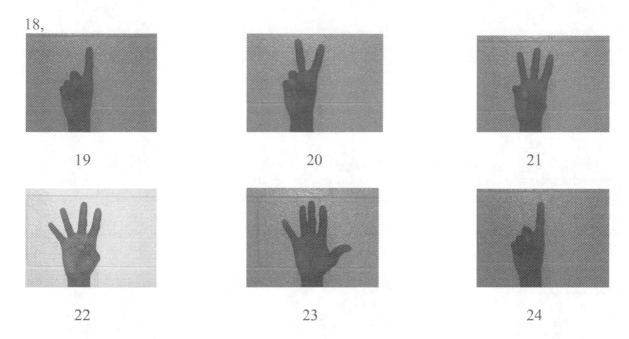

| 19 | 20 | 21 |
| 22 | 23 | 24 |

Students have counted up 6 spaces from the product of $6 \times 3 = 18$, so they know that $6 \times 4 = 24$. Students manually list the new product.

$6 \times 1 = 6$

$6 \times 2 = 12$

$6 \times 3 = 18$

$6 \times 4 = 24$

Now students begin the process again to determine the product of the next factor, which is 5. To determine the new product, students will count up 6 spaces from the product of 24. Students will raise one finger as they count up each space.

$6 \times 5 =$

24,

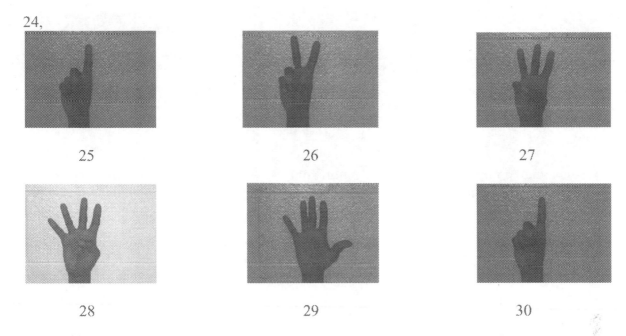

25

26

27

28

29

30

Students have counted up 6 spaces from the product of 6 × 4 = 24, so they know that 6 × 5 = 30. Students manually list the new product.

6 × 1 = 6

6 × 2 = 12

6 × 3 = 18

6 × 4 = 24

6 × 4 = 30

Now students begin the process again to determine the product of the next factor, which is 6. To determine the new product, students will count up 6 spaces from the product of 30. Students will raise one finger as they count up each space.

6 × 6 =

30,

31

32

33

34   35   36

Students have counted up 6 spaces from the product of 6 × 5 = 30, so they know that 6 × 6 = 36. Students manually list the new product.

6 × 1 = 6   6 × 5 = 30

6 × 2 = 12   6 × 6 = 36

6 × 3 = 18

6 × 4 = 24

Now students begin the process again to determine the product of the next factor, which is 7. To determine the new product, students will count up 6 spaces from the product of 36. Students will raise one finger as they count up each space.

6 × 7 =

36,

37   38   39

40   41   42

Students have counted up 6 spaces from the product of 6 × 6 = 36, so they know that 6 × 7 = 42. Students manually list the new product.

$6 \times 1 = 6$ $6 \times 5 = 30$

$6 \times 2 = 12$ $6 \times 6 = 36$

$6 \times 3 = 18$ $6 \times 7 = 42$

$6 \times 4 = 24$

At this point, help students recognize that the target number of 42 has been identified. $6 \times 7 = 42$, so 6 will divide into 42, **7 times**. Students will manually write out the entire step.

Now explain to students the entire process: $6 \times 7 = 42$; to determine the remainder, subtract the product from the target number. $42 - 42$ is equal to 0 for a remainder.

It is very important to model this step on the chalkboard and with students individually until they grasp the concept

$$\begin{array}{r} 7 \\ \hline 6\overline{)42} \\ -42 \\ \hline 0\ R \end{array}$$

Students will use scratch paper to manually write out the table of 6 as they identify each new product until the target number is reached. Once the target number is reached, students will manually write out each step in the process of solving the long-division equation. The classroom teacher should identify mistakes and areas of weakness by reviewing the scratch paper and reviewing concepts as necessary.

Long Division: Table of 8

$72 \div 8 =$

Step 1: Help students determine the target number for the equation. The target number in this equation is 72.

Step 2: Students determine which multiplication table will be used to solve the equation. In this equation, the table of 8 will be used, because it must be determined how many times 8 will divide into 72.

Step 3: Students write out the first multiple of the table of 8, which will be 1. (This process of writing out the table of 8 will continue until the missing multiple is identified that will produce a product of 72 when multiplied by 8.)

$8 \times 1 = 8$

Step 4: Students apply the strategy of counting up using their fingers to help them determine the next product, of 8 times 2.

$8 \times 2 =$

8,

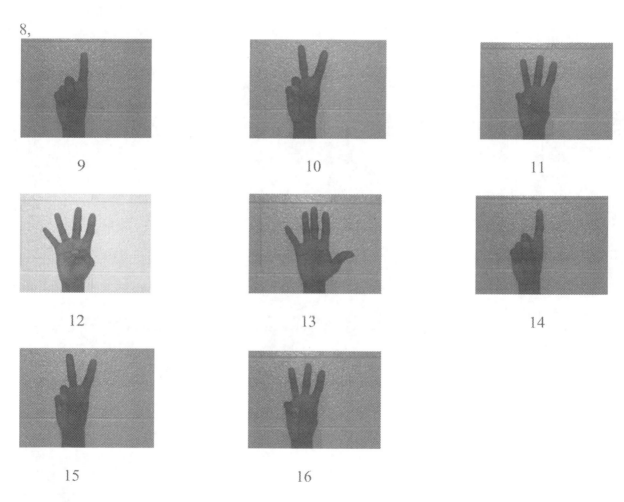

9	10	11
12	13	14
15	16	

Students have counted up 8 spaces from the product of $8 \times 1 = 8$, so they know that $8 \times 2 = 16$. Students manually list the new product.

$8 \times 1 = 8$

$8 \times 2 = 16$

Now students begin the process again to determine the product of the next factor, which is 3. To determine the new product, students count up 8 spaces from the product of 16. Students will raise one finger as they count up each space.

$8 \times 3 =$

16,

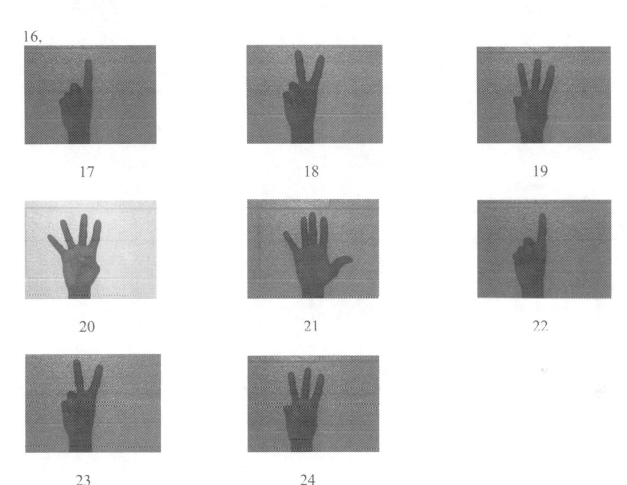

17   18   19

20   21   22

23   24

Students have counted up 8 spaces from the product of $8 \times 2 = 16$, so they know that $8 \times 3 = 24$. Students manually list the new product.

$8 \times 1 = 8$

$8 \times 2 = 16$

$8 \times 3 = 24$

Now students begin the process again to determine the product of the next factor, which is 4. To determine the new product, students count up 8 spaces from the product of 24. Students will raise one finger as they count up each space.

$8 \times 4 =$

24,

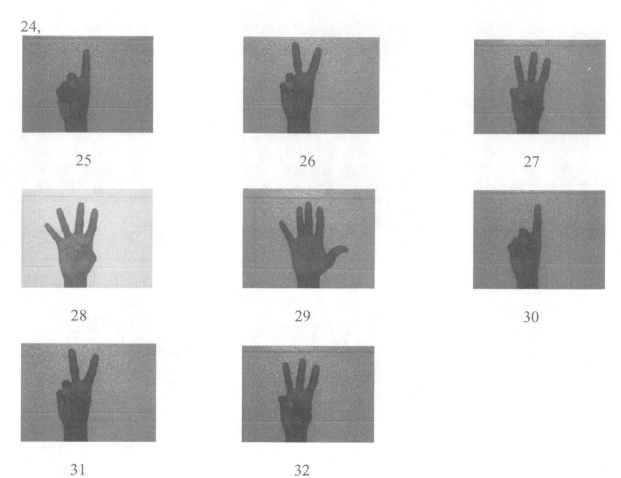

25

26

27

28

29

30

31

32

Students have counted up 8 spaces from the product of 8 × 3 = 24, so they know that 8 × 4 = 32.
Students manually list the new product.

8 × 1 = 8

8 × 2 = 16

8 × 3 = 24

8 × 4 = 32

Now students begin the process again to determine the product of the next factor, which is 5. To
determine the new product, students count up 8 spaces from the product of 32. Students will raise
one finger as they count up each space.

8 × 5 =

32,

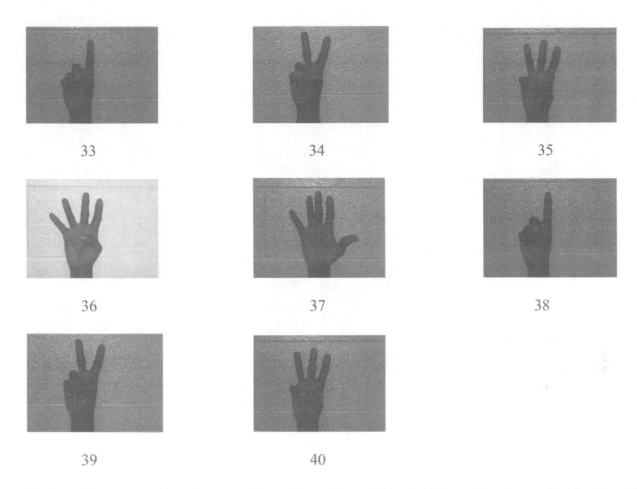

33 34 35

36 37 38

39 40

Students have counted up 8 spaces from the product of 8 × 4 = 32, so they know that 8 × 5 = 40.
Students manually list the new product.

8 × 1 = 8   8 × 5 = 40

8 × 2 = 16

8 × 3 = 24

8 × 4 = 32

Now students begin the process again to determine the product of the next factor, which is 6. To
determine the new product, students count up 8 spaces from the product of 40. Students will raise
one finger as they count up each space.

8 × 6 =

40,

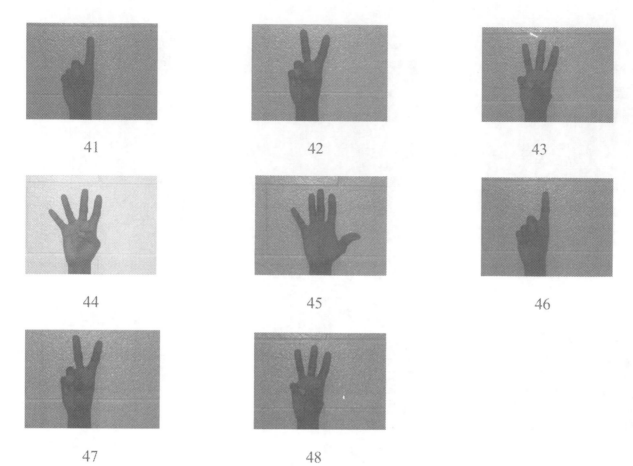

41   42   43

44   45   46

47   48

Students have counted up 8 spaces from the product of $8 \times 5 = 40$, so they know that $8 \times 6 = 48$. Students manually list the new product.

$8 \times 1 = 8$ $8 \times 5 = 40$

$8 \times 2 = 16$ $8 \times 6 = 48$

$8 \times 3 = 24$

$8 \times 4 = 32$

Now students begin the process again to determine the product of the next factor, which is 7. To determine the new product, students count up 8 spaces from the product of 48. Students will raise one finger as they count up each space.

$8 \times 7 =$

48,

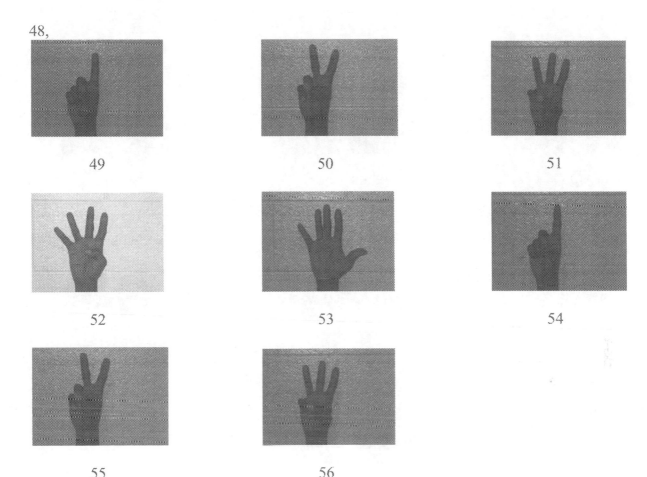

49   50   51

52   53   54

55   56

Students have counted up 8 spaces from the product of 8 × 6 = 48, so they know that 8 × 7 = 56. Students manually list the new product.

8 × 1 = 8   8 × 5 = 40

8 × 2 = 16   8 × 6 = 48

8 × 3 = 24   8 × 7 = 56

8 × 4 = 32

Now students begin the process again to determine the product of the next factor, which is 8. To determine the new product, students count up 8 spaces from the product of 56. Students will raise one finger as they count up each space

8 × 8 =

56,

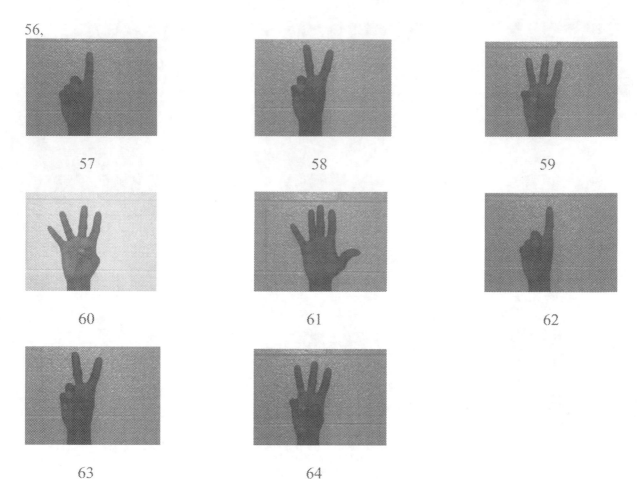

57

58

59

60

61

62

63

64

Students have counted up 8 spaces from the product of 8 × 7 = 56, so they know that 8 × 8 = 64, Students manually list the new product.

8 × 1 = 8   8 × 6 = 48

8 × 2 = 16   8 × 7 = 56

8 × 3 = 24   8 × 8 = 64

8 × 4 = 32

8 × 5 = 40

Now students begin the process again to determine the product of the next factor, which is 9. To determine the new product, students count up 8 spaces from the product of 64. Students will raise one finger as they count up each space.

8 × 9 =

64,

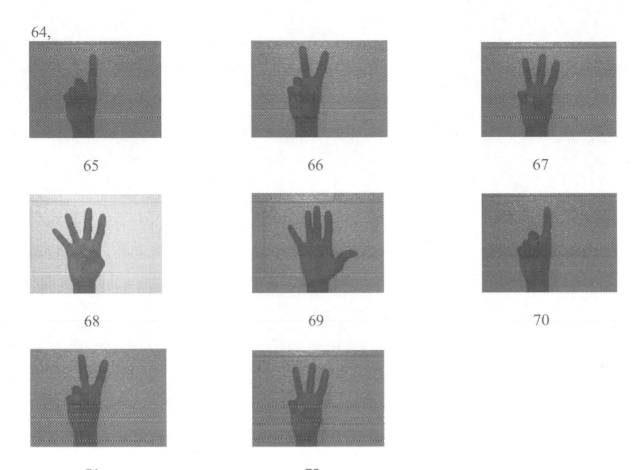

65

66

67

68

69

70

71

72

Students have counted up 8 spaces from the product of 8 × 8 = 64, so they know that 8 × 9 = 72. Students manually list the new product.

8 × 1 = 8   8 × 6 = 48

8 × 2 = 16   8 × 7 = 56

8 × 3 = 24   8 × 8 = 64

8 × 4 = 32   8 × 9 = 72

8 × 5 = 40

At this point, the target number of 72 has been identified. 8 × 9 = 72, so 72 ÷ 8 = 9.

Chapter 4
Inverse Operation

The inverse operation of division

$42 \div 6 = 7$

is multiplication:

$6 \times 7 = 42$

The Rule of Dividing by Zero:
Any number divided by 0 is equal to 0.
Example: $42 \div 0 = 0$

Compared with the traditional teaching method, we will place more emphasis on division problems. Through the application of division, students learn multiplication, division, and the time tables simultaneously. Students will begin with a random selection of division problems. Be sure to use a variety of problems that include all multiplication factors varying from 1 to 12.

As students work to solve each division problem, it is important that they manually write out each sequence of the time tables that is being used to count up to until the target number or product is identified. Initially this step is slow paced and time consuming, but the rewards are worth the added effort. The repetition of writing out the time tables as students count up to solve division problems will help students learn multiplication, division, and the time table chart all at the same time. Always have students begin multiplying using the first multiple of 1 as they manually write out the table.

In this example, the table of 6 will be used.

$42 \div 6 =$

$6 \times 1 = 6$ $6 \times 4 = 24$ $6 \times 7 = 42$

$6 \times 2 = 12$ $6 \times 5 = 30$

$6 \times 3 = 18$ $6 \times 6 = 36$

Another reason students manually write out the time table of the multiple used is to help you identify areas of weakness that need additional practice. The students' scratch paper can be collected for this purpose.

Chapter 5
Division Problems and Counting-Up Strategy

$42 \div 6 =$ $72 \div 8 =$ $49 \div 7 =$ $4 \div 2 =$

$3 \div 3 =$ $50 \div 10 =$ $48 \div 12 =$ $66 \div 11 =$

During the beginning stages of teaching students the strategy of counting up, close facilitations, whole-group instruction, direct instruction, and one-on-one modeling are required until students grasp the concept.

Before your students begin counting up, it is very important to help students identify the target number.

$42 \div 6 =$

Identify the target number by making a clear statement: "Our target number in this equation is 42." Explain that students should now try to determine the number of times 6 will divide into 42.

To determine the number of times 6 will divide into 42, we will list the time table of 6 until we find the number that, when multiplied with 6, will produce the product of 42.

Counting Up Using the Time Table of 6

Step 1: Tell students to remember the rule for multiplying by 1, which is that any number multiplied by 1 equals that number.

$6 \times 1 = 6$
Step 2: Explain to students that this is where they will count up to identify the product of 6×2. Since they are multiplying using the table of 6, they should count up 6 spaces from 6 (which is the product of 6×1).

$6 \times 2 =$

Begin counting aloud with the number 6. Next, count aloud 7 and raise a finger; count aloud 8 and raise the second finger; count aloud 9 and raise the third finger; count aloud 10 and raise the fourth finger; count aloud 11 and raise the fifth finger; and count aloud 12 and raise the sixth finger.

 6,

Count up 1 space: 7   Count up 2 spaces: 8   Count up 3 spaces: 9

Count up 4 spaces: 10

Count up 5 spaces: 11

Count up 6 spaces: 12

When finished, you will have one hand with five fingers raised; the second hand will have only one finger raised. You have counted up 6 spaces to the number 12; the product of 6 × 2 is 12.

6 × 2 = 12

Step 3: Explain to students that they now know that 6 × 2 = 12, so now they can determine the product of 6 × 3.

6 × 3 =

Begin counting aloud with the number 12. Next, count aloud 13 and raise one finger; count aloud 14 and raise the second finger; count aloud 15 and raise the third finger; count aloud 16 and raise the fourth finger; count aloud 17 and raise the fifth finger; and count aloud 18 and raise the sixth finger.

12,

Count up 1 space: 13

Count up 2 spaces: 14

Count up 3 spaces: 15

Count up 4 spaces: 16

Count up 5 spaces: 17

Count up 6 spaces: 18

When finished, you will have one hand with five fingers raised; the second hand will have only one finger raised. You have counted up 6 spaces to the number 18; the product of 6 × 3 is 18.

6 × 3 = 18

Step 4: Explain to students that they now know that 6 × 3 = 18, so now they can determine the product of 6 × 4.
6 × 4 =

Begin by counting aloud 18. Next, count aloud 19 and raise one finger; count aloud 20 and raise the second finger; count aloud 21 and raise the third finger; count aloud 22 and raise the fourth finger; count aloud 23 and raise the fifth finger; count aloud 24 and raise the sixth finger.
18,

Count up 1 space: 19   Count up 2 spaces: 20   Count up 3 spaces: 21

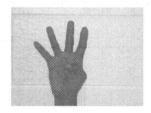

Count up 4 spaces: 22   Count up 5 spaces: 23   Count up 6 spaces: 24

When finished, you will have one hand with five fingers raised; the second hand will have only one finger raised. You have counted up 6 spaces to the product of 24; the product of 6 × 4 is 24.

6 × 4 = 24
Step 5: Explain to students that they now know that 6 × 4 = 24, so now they can determine the product of 6 × 5.

6 × 5 =
Begin counting with the number 24. Next, count aloud 25 and raise one finger; count aloud 26 and raise the second finger; count aloud 27 and raise the third finger; count aloud 28 and raise the fourth finger; count aloud 29 and raise the fifth finger; count aloud 30 and raise the sixth finger.
24,

Count up 1 space: 25   Count up 2 spaces: 26   Count up 3 spaces: 27

Count up 4 spaces: 28   Count up 5 spaces: 29   Count up 6 spaces: 30

When finished, you will have one hand with five fingers raised; the second hand will have only one finger raised. You have counted up 6 spaces to the number 30; the product of 6 × 5 is 30.

6 × 5 = 30

Step 6: Explain to students that they now know that 6 × 5 = 30, so now they can determine the product of 6 × 6.

6 × 6 =

Begin by counting aloud 30. Next, count aloud 31 and raise one finger; count aloud 32 and raise the second finger; count aloud 33 and raise the third finger; count aloud 34 raise the fourth finger; count aloud 35 and raise the fifth finger; and count aloud 36 and raise the sixth finger. 30,

Count up 1 space: 31   Count up 2 spaces: 32   Count up 3 spaces: 33

Count up 4 spaces: 34   Count up 5 spaces: 35   Count up 6 spaces: 36

When finished, you will have one hand with five fingers raised; the second hand will have only one finger raised. You have counted up 6 spaces to the number 36; the product of 6 × 6 is 36.

Count up a total of 6 spaces from the product 30
6 × 6 = 36

Step 7: Explain to students that they now know that 6 × 6 = 36, so now they can determine the product of 6 × 7.

6 × 7 =

Begin by counting aloud 36. Next, count aloud 37 and raise one finger; count aloud 38 and raise the second finger; count aloud 39 and raise the third finger; count aloud 40 and raise the fourth finger; count aloud 41 and raise the fifth finger; count aloud 42 and raise the sixth finger.
36,

Count up 1 space: 37

Count up 2 spaces: 38

Count up 3 spaces: 39

Count up 4 spaces: 40

Count up 5 spaces: 41

Count up 6 spaces: 42

When finished, you will have one hand with five fingers raised; the second hand will have only one finger raised. You have counted up 6 spaces to the number 42; the product of 6 × 7 is 42.

6 × 7 = 42

At this point, remind students that we are trying to determine the number of times 6 will divide into 42.

Point out that if 6 × 7 = 42, 42 ÷ 6 = 7, so 6 will divide into 42, **7 times**. The target number of 42 has been identified.

Counting Up Using the Table of 8

Our next equation is

$72 \div 8 = ?$

Ask students to identify the target number. Wait for responses, and then coax students by stating, "We are looking for 72. What number when multiplied with 8 is equal to 72?" Explain that this time we will be using the time table of 8, so we will be counting up 8 spaces beginning with the product of 8×1.

Step 1: Tell students to remember the rule for multiplying by 1: any number multiplied by 1 equals that number.
$8 \times 1 = 8$

Step 2: Explain to students that this is where they will begin to count up to determine the product of 8×2.

$8 \times 2 = ?$
Begin counting aloud with the number 8. Next, count aloud 9 and raise one finger; count aloud 10 and raise the second finger; count aloud 11 and raise the third finger; count aloud 12 and raise the fourth finger; count aloud 13 and raise the fifth finger; count aloud 14 and raise the sixth finger; count aloud 15 and raise the seventh finger; count aloud 16 and raise the eighth finger.
8,

Count up 1 space: 9

Count up 2 spaces: 10

Count up 3 spaces: 11

Count up 4 spaces: 12

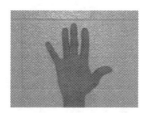

Count up 5 spaces: 13

Count up 6 spaces: 14

Count up 7 spaces: 15   Count up 8 spaces: 16

When finished, you will have one hand with five fingers raised; the second hand will have only three fingers raised. You have counted up 8 spaces to the number 16; the product of 8 × 2 is 16.

Count up a total of 8 spaces
8 × 2 = 16

Step 3: Explain to students that they now know that 8 × 2 = 16, so now they can determine the product of 8 × 3.

8 × 3 =?

Begin counting aloud with the number 16. Next, count aloud 17 and raise one finger; count aloud 18 and raise the second finger; count aloud 19 and raise the third finger; count aloud 20 and raise the fourth finger; count aloud 21 and raise the fifth finger; count aloud 22 and raise the sixth finger; count aloud 23 and raise the seventh finger; and count aloud 24 and raise the eighth finger.

16,

Count up 1 space: 17   Count up 2 spaces: 18   Count up 3 spaces: 19

Count up 4 spaces: 20   Count up 5 spaces: 21   Count up 6 spaces: 22

Count up 7 spaces: 23   Count up 8 spaces: 24

When finished, you will have one hand with five fingers raised; the second hand will have only three fingers raised. You have counted up 8 spaces to the number 24; the product of 8 × 3 is 24.

Count up a total of 8 spaces
8 × 3 = 24

Step 4: Explain to students that they now know that the product of 8 × 3 = 24, so now they can determine the product of 8 × 4.

8 × 4 =?

Begin counting aloud with the number 24. Next, count aloud 25 and raise one finger; count aloud 26 and raise the second finger; count aloud 27 and raise the third finger; count aloud 28 and raise the fourth finger; count aloud 29 and raise the fifth finger; count aloud 30 and raise the sixth finger; count aloud 31 and raise the seventh finger; and count aloud 32 and raise the eighth finger.
24,

Count up 1 space: 25   Count up 2 spaces: 26   Count up 3 spaces: 27

Count up 4 spaces: 28   Count up 5 spaces: 29   Count up 6 spaces: 30

Count up 7 spaces: 31   Count up 8 spaces: 32

When finished, you will have one hand with five fingers raised; the second hand will have only three fingers raised. You have counted up 8 spaces to the number 32; the product of 8 × 4 is 32. Count up a total of 8 spaces
8 × 4 = 32

Step 5: Explain to students that they now know that 8 × 4 = 32, so now they can determine the product 8 × 5.

8 × 5 =?

Keep in mind we are looking for the target number of 72.

Begin counting aloud with the number 32. Next, count aloud 33 and raise one finger; count aloud 34 and raise the second finger; count aloud 35 and raise the third finger; count aloud 36 and raise the fourth finger; count aloud 37 and raise the fifth finger; count aloud 38 and raise the sixth finger; count aloud 39 and raise the seventh finger; and count aloud 40 and raise the eighth finger.
32,

Count up 1 space: 33   Count up 2 spaces: 34   Count up 3 spaces: 35

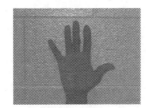

Count up 4 spaces: 36   Count up 5 spaces: 37   Count up 6 spaces: 38

Count up 7 spaces: 39   Count up 8 spaces: 40

When finished, you will have one hand with five fingers raised; the second hand will have only three fingers raised. You have counted up 8 spaces to the number 40; the product of 8 × 5 is 40.

Count up a total of 8 spaces
8 × 5 = 40

Step 6: Explain to students that they now know that 8 × 5 = 40, so now they can determine the product of 8 × 6.

8 × 6 =?

Keep in mind we are looking for the target number of 72.

Begin counting aloud with the number 40. Next, count aloud 41 and raise one finger; count aloud 42 and raise the second finger; count aloud 43 and raise the third finger; count aloud 44 and raise the fourth finger; count aloud 45 and raise the fifth finger; count aloud 46 and raise the sixth finger; count aloud 47 and raise the seventh finger; and count aloud 48 and raise the eighth finger.
40,

Count up 1 space: 41   Count up 2 spaces: 42   Count up 3 spaces: 43

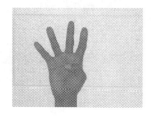

Count up 4 spaces: 44   Count up 5 spaces: 45   Count up 6 spaces: 46

Count up 7 spaces: 47

Count up 8 spaces: 48

When finished, you will have one hand with five fingers raised; the second hand will have only three fingers raised. You have counted up 8 spaces to the number 48; the product of 8 × 6 is 48.

Count up a total of 8 spaces
8 × 6 = 48

Step 7: Explain to students that they now know that 8 × 6 = 48, so now they can determine the product of 8 × 7.

8 × 7 = ?

Begin counting aloud with the number 48. Next, count aloud 49 and raise one finger; count aloud 50 and raise the second finger; count aloud 51 and raise the third finger; count aloud 52 and raise the fourth finger; count aloud 53 and raise the fifth finger; count aloud 54 and raise the sixth finger; count aloud 55 and raise the seventh finger; and count aloud 56 and raise the eighth finger.

48,

Count up 1 space: 49

Count up 2 spaces: 50

Count up 3 spaces: 51

Count up 4 spaces: 52

Count up 5 spaces: 53

Count up 6 spaces: 54

Count up 7 spaces: 55   Count up 8 spaces: 56

When finished, you will have one hand with five fingers raised; the second hand will have only three fingers raised. You have counted up 8 spaces to the number 56; the product of 8 × 7 is 56.

Count up a total of 8 spaces
8 × 7 = 56

Step 8: Explain to students that they now know that the product of 8 × 7 = 56, so now they can determine the product of 8 × 8.

8 × 8 =?

Keep in mind we are looking for the target number of 72.

Begin counting aloud with the number 56. Next, count aloud 57 and raise one finger; count aloud 58 and raise the second finger; count aloud 59 and raise the third finger; count aloud 60 and raise the fourth finger; count aloud 61 and raise the fifth finger; count aloud 62 and raise the sixth finger; count aloud 63 and raise the seventh finger; count aloud 64 and raise the eighth finger.
   56,

Count up 1 space: 57   Count up 2 spaces: 58   Count up 3 spaces: 59

Count up 4 spaces: 60   Count up 5 spaces: 61   Count up 6 spaces: 62

42

Count up 7 spaces: 63   Count up 8 spaces: 64

When finished, you will have one hand with five fingers raised; the second hand will have only three fingers raised. You have counted up 8 spaces to the number 64; the product of 8 × 8 is 64.

Count up a total of 8 spaces
8 × 8 = 64

Step 9: Explain to students that they now know that 8 × 8 = 64, so now they can determine the product of 8 × 9.

8 × 9 =?

Keep in mind we are looking for the target number of 72.

Begin counting aloud with the number 64. Next, count aloud 65 and then raise one finger; count aloud 66 and raise the second finger; count aloud 67 and raise the third finger; count aloud 68 and raise the fourth finger; count aloud 69 and raise the fifth finger; count aloud 70 and raise the sixth finger; count aloud 71 and raise the seventh finger; and count aloud 72 and raise the eighth finger.
64,

Count up 1 space: 65   Count up 2 spaces: 66   Count up 3 spaces: 67

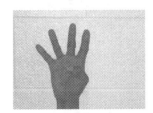

Count up 4 spaces: 68   Count up 5 spaces: 69   Count up 6 spaces: 70

43

Count up 7 spaces: 71   Count up 8 spaces: 72

When finished, you will have one hand with five fingers raised; the second hand will have only three fingers raised. You have counted up 8 spaces to the number 72; the product of 8 × 9 is 72.

Count up a total of 8 spaces
8 × 9 = 72
Target number = 72

During each step, as students apply the strategy of counting up, have students manually write out each of the two factors, along with its product, until the target number is reached. The example below illustrates how students will write out the multiples of the table of 8. Remember to have students begin with 1 as the first multiple.

8 × 1 = 8   8 × 5 = 40   8 × 9 = 72

8 × 2 = 16   8 × 6 = 48

8 × 3 = 24   8 × 7 = 56

8 × 4 − 32   8 × 8 = 64

Note once the target number is reached, students stop and move on to the next division or multiplication equation.

Counting Up Using the Table of 7
Our next equation is

49 ÷ 7 =?
In this example, the table of 7 is used. Students will begin writing out the table of 7 beginning with the multiple of 1; continue the sequence after each product is identified until the target number is reached. Note the example below. Identify the product, write the time table sequence.

7 × 1 = 7   7 × 4 = 28   7 × 7 = 49

7 × 2 = 14   7 × 5 = 35

7 × 3 = 21   7 × 6 = 42

Identify the target number students will search for by making a clear statement: "The target number we will be searching for in this equation is 49. Explain that this time, we will be using the table of 7, so we will be counting up 7 spaces, beginning with 7 (the product of 7×1).

Step 1: Tell students to remember the rule for multiplying by 1: any number multiplied by 1 equals that number.

$7 \times 1 = 7$

Step 2: Explain to students that this is where they will begin to count up to determine the product of $7 \times 2 = ?$

Begin counting up to determine the product of 7×2.

Begin counting aloud with the number 7. Next, count aloud 8 and raise one finger; count aloud 9 and raise the second finger; count aloud 10 and raise the third finger; count aloud 11 and raise the fourth finger; count aloud 12 and raise the fifth finger; count aloud 13 and raise the sixth finger; and count aloud 14 and raise the seventh finger.

7,

Count up 1 space: 8

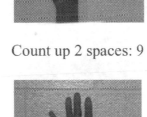

Count up 2 spaces: 9

Count up 3 spaces: 10

Count up 4 spaces: 11

Count up 5 spaces: 12

Count up 6 spaces: 13

Count up 7 spaces: 14

45

When finished, you will have one hand with five fingers raised; the second hand will have only two fingers raised. You have counted up 7 spaces to the number 14; the product of 7 × 2 is 14.

Count up a total of 7 spaces
7 × 2 = 14

Step 3: Explain to students that they now know that 7 × 2 = 14, so now they can determine the product of 7 × 3.

7 × 3 =?

Begin counting aloud with the number 14. Next, count aloud 15 and raise one finger; count aloud 16 and raise the second finger; count aloud 17 and raise the third finger; count aloud 18 and raise the fourth finger; count aloud 19 and raise the fifth finger; count aloud 20 and raise the sixth finger; count aloud 21 and raise the seventh finger.
14,

Count up 1 space: 15

Count up 2 spaces: 16

Count up 3 spaces: 17

Count up 4 spaces: 18

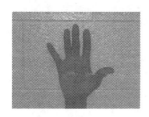

Count up 5 spaces: 19

Count up 6 spaces: 20

Count up 7 spaces: 21

When finished, you will have one hand with five fingers raised; the second hand will have only two fingers raised. You have counted up 7 spaces to the number 21; the product of 7 × 3 is 21.

Count up a total of 7 spaces

Step 4: Explain to students that they now know that $7 \times 3 = 21$, so now they can determine the product of 7×4.

$7 \times 4 = ?$

Begin counting aloud beginning with 21. Next, count aloud 22 and raise one finger; count aloud 23 and raise the second finger; count aloud 24 and raise the third finger; count aloud 25 and raise the fourth finger; count aloud 26 and raise the fifth finger; count aloud 27 and raise the sixth finger; count aloud 28 and raise the seventh finger.
21,

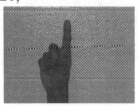
Count up 1 space: 22

Count up 2 spaces: 23

Count up 3 spaces: 24

Count up 4 spaces: 25

Count up 5 spaces: 26

Count up 6 spaces: 27

Count up 7 spaces: 28

When finished, you will have one hand with five fingers raised; the second hand will have only two fingers raised. You have counted up 7 spaces to the number 28; the product of 7×4 is 28.

Count up a total of 7 spaces
$7 \times 4 = 28$

Step 5: Explain to students that they now know that $7 \times 4 = 28$, so now they can determine the product of 7×5.

$7 \times 5 =?$

Begin counting aloud with the number 28. Next, count aloud 29 and raise one finger; count aloud 30 and raise the second finger; count aloud 31 and raise the third finger; count aloud 32 and raise the fourth finger; count aloud 33 and raise the fifth finger; count aloud 34 and raise the sixth finger; and count aloud 35 and raise the seventh finger.

28,

Count up 1 space: 29

Count up 2 spaces: 30

Count up 3 spaces: 31

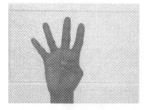

Count up 4 spaces: 32

Count up 5 spaces: 33

Count up 6 spaces: 34

Count up 7 spaces: 35

When finished, you will have one hand with five fingers raised; the second hand will have only two fingers raised. You have counted up 7 spaces to the number 35; the product of 7×5 is 35.

Count up a total of 7 spaces
$7 \times 5 = 35$

Step 3: Explain to students that they now know that $7 \times 5 = 35$, so now they can determine the product of 7×6.

$7 \times 6 =?$

Begin counting aloud with the number 35. Next, count aloud 36 and raise one finger; count aloud 37 and raise the second finger; count aloud 38 and raise the third finger; count aloud 39 and raise the fourth finger; count aloud 40 and raise the fifth finger; count aloud 41 and raise the sixth finger; and count aloud 42 and raise the seventh finger.

35,

Count up 1 space: 36

Count up 2 spaces: 37

Count up 3 spaces: 38

Count up 4 spaces: 39

Count up 5 spaces: 40

Count up 6 spaces: 41

Count up 7 spaces: 42

When finished, you will have one hand with five fingers raised; the second hand will have only two fingers raised. You have counted up 7 spaces to the number 42; the product of 7 × 6 is 42.

Count up a total of 7 spaces
7 × 6 = 42

Step 3: Explain to students that they now know that 7 × 6 = 42, so now they can determine the product of 7 × 7.

7 × 7 = ?

Begin counting aloud with the number 42. Next, count aloud 43 and raise one finger; count aloud 44 and raise the second finger; count aloud 45 and raise the third finger; count aloud 46 and raise the fourth finger; count aloud 47 and raise the fifth finger; count aloud 48 and raise the sixth finger; and count aloud 49 and raise the seventh finger.

42,

Count up 1 space: 43   Count up 2 spaces: 44   Count up 3 spaces: 45

Count up 4 spaces: 46   Count up 5 spaces: 47   Count up 6 spaces: 48

Count up 7 spaces: 49

When finished, you will have one hand with five fingers raised; the second hand will have only two fingers raised. You have counted up 7 spaces to the number 49; the product of 7×7 is 49.

Count up a total of 7 spaces
$7 \times 7 = 49$
Target number = 49
$49 \div 7 = 7$

Counting Up Using the Time Table of 2

Our next equation is

$4 \div 2 =?$

Identify the target number students will be searching for by stating, "Our target number in this equation is 4." Explain that we are trying to determine the number of time 2 will divide into 4.

To determine the number of times 2 will divide into 4, we will list the time table of 2 until we find the number that, when multiplied with 2, is equal to 4.

the fourth finger; count aloud 40 and raise the fifth finger; count aloud 41 and raise the sixth finger; and count aloud 42 and raise the seventh finger.

35,

Count up 1 space: 36

Count up 2 spaces: 37

Count up 3 spaces: 38

Count up 4 spaces: 39

Count up 5 spaces: 40

Count up 6 spaces: 41

Count up 7 spaces: 42

When finished, you will have one hand with five fingers raised; the second hand will have only two fingers raised. You have counted up 7 spaces to the number 42; the product of 7×6 is 42.

Count up a total of 7 spaces
$7 \times 6 = 42$

Step 3: Explain to students that they now know that $7 \times 6 = 42$, so now they can determine the product of 7×7.

$7 \times 7 = ?$

Begin counting aloud with the number 42. Next, count aloud 43 and raise one finger; count aloud 44 and raise the second finger; count aloud 45 and raise the third finger; count aloud 46 and raise the fourth finger; count aloud 47 and raise the fifth finger; count aloud 48 and raise the sixth finger; and count aloud 49 and raise the seventh finger.

42,

Count up 1 space: 43

Count up 2 spaces: 44

Count up 3 spaces: 45

Count up 4 spaces: 46

Count up 5 spaces: 47

Count up 6 spaces: 48

Count up 7 spaces: 49

When finished, you will have one hand with five fingers raised; the second hand will have only two fingers raised. You have counted up 7 spaces to the number 49; the product of 7 × 7 is 49.

Count up a total of 7 spaces
7 × 7 = 49
Target number = 49
49 ÷ 7 = 7

Counting Up Using the Time Table of 2

Our next equation is

4 ÷ 2 =?

Identify the target number students will be searching for by stating, "Our target number in this equation is 4." Explain that we are trying to determine the number of time 2 will divide into 4.

To determine the number of times 2 will divide into 4, we will list the time table of 2 until we find the number that, when multiplied with 2, is equal to 4.

Step 1: Tell students to remember the rule for multiplying by 1: any number multiplied by 1 equals that number.

$2 \times 1 = 2$

Step 2: Explain to students that this is where they will begin to count up to identify the product of 2×2. Since we are multiplying using the table of 2, 2 should be added to the product of 2×1, which is 2. So start counting up 2 spaces from the number 2.

$2 \times 2 = ?$

Begin counting aloud with the number 2. Next, count aloud 3 and raise one finger; count aloud 4 and raise the second finger.

2,

Count up 1 space: 3   Count up 2 spaces: 4

Count up a total of 2 spaces
$2 \times 2 = 4$
Target number = 4
$4 \div 2 = 2$

In this example, the table of 2 will be used. Students will begin writing out the table of 2 beginning with the number 1 and continue until the target number is reached.

$2 \times 1 = 2$

$2 \times 2 = 4$

Counting Up Using the Time Table of 3

Our next equation is

$3 \div 3 = ?$

Identify the target number students will be searching for. The target number in this equation is 3. Explain that we are trying to determine the number of times 3 will divide into 3.

To determine the number of times 3 will divide into 3, we will list the time table of 3 until we find the number that, when multiplied with 3, is equal to 3.

Step 1: Tell students to remember the rule for multiplying by 1: any number multiplied by 1 equals that number.

$3 \times 1 = 3$
Target number = 3

Explain to students that in this equation, the target number is 3, so there is no need to go any further than Step 1; only the first step is required in solving this problem.

$3 \times 1 = 3$
Target number = 3
$3 \div 3 = 1$

In this example students will begin writing the table of 3, beginning with 1 and continuing until the target number is reached.

$3 \times 1 = 3$

It is important to review the rules of multiplying by one and dividing a number by itself.

Counting Up Using the Time Table of 10

Our next equation is

$50 \div 10 = ?$

Identify the target number students will be searching for by making a clear statement: "Our target number in this equation is 50." Explain that we are trying to determine the number of times 10 will divide into 50.

To determine the number of times 10 will divide into 50, we will list the time table of 10 until we find the number that, when multiplied with 10, is equal to 50.

Step 1: Tell students to remember the rule for multiplying by 1: any number multiplied by 1 equals that number.

$10 \times 1 = 10$

Step 2: Explain to students that this is where they will begin to count up to determine the product of $10 \times 2 = ?$

Begin counting aloud with the number 10. Next, count aloud to 11 and raise one finger; count aloud to 12 and raise the second finger; count aloud 13 and raise the third finger; count aloud 14 and raise the fourth finger; count aloud 15 and raise the fifth finger; count aloud 16 and raise the sixth finger; count aloud 17 and raise the seventh finger; count 18 and raise the eighth finger; count aloud 19 and raise the ninth finger; and count aloud 20 and raise the tenth finger.

10,

Count up 1 space: 11   Count up 2 spaces: 12   Count up 3 spaces: 13

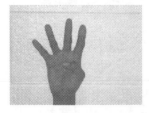  

Count up 4 spaces: 14   Count up 5 spaces: 15   Count up 6 spaces: 16

Count up 7 spaces: 17   Count up 8 spaces: 18   Count up 9 spaces: 19

Count up 10 spaces: 20

Count up a total of 10 spaces
$10 \times 2 = 20$

Step 3: Explain to students that they now know that $10 \times 2 = 20$, so now they can determine the product of 10×3

$10 \times 3 = ?$

Begin counting aloud with the number 20. Next, count aloud 21 and raise one finger; count aloud 22 and raise the second finger; count aloud 23 and raise the third finger; count aloud 24 and raise the fourth finger; count aloud 25 and raise the fifth finger; count aloud 26 and raise the sixth finger; count aloud 27 and raise the seventh finger; count 28 and raise the eighth finger; count aloud 29 and raise the ninth finger; count aloud 30 and raise the tenth finger.

20,

Count up 1 space: 21

Count up 2 spaces: 22

Count up 3 spaces: 23

Count up 4 spaces: 24

Count up 5 spaces: 25

Count up 6 spaces: 26

Count up 7 spaces: 27

Count up 8 spaces: 28

Count up 9 spaces: 29

Count up 10 spaces: 30

Count up a total of 10 spaces
10 × 3 = 30

Step 4: Explain to students that they now know that 10 × 3 = 30, so now they can determine the product of 10 × 4.

10 × 4 = ?

54

Begin counting aloud with the number 30. Next, count aloud 31 and raise one finger; count aloud 32 and raise the second finger; count aloud 33 and raise the third finger; count aloud 34 and raise the fourth finger; count aloud 35 and raise the fifth finger; count aloud 36 and raise the sixth finger; count aloud 37 and raise the seventh finger; count aloud 38 and raise the eighth finger; count aloud 39 and raise the ninth finger; and count aloud 40 and raise the tenth finger.

30,

Count up 1 space: 31

Count up 2 spaces: 32

Count up 3 spaces: 33

Count up 4 spaces: 34

Count up 5 spaces: 35

Count up 6 spaces: 36

Count up 7 spaces: 37

Count up 8 spaces: 38

Count up 9 spaces: 39

Count up 10 spaces: 40

Count up a total of 10 spaces

$10 \times 4 = 40$

Step 5: Explain to students that they now know that $10 \times 4 = 40$, so now they can determine the product of 10×5.

$10 \times 5 = ?$

Begin counting aloud with the number 40. Next, count aloud 41 and raise one finger; count aloud 42 and raise the second finger; count aloud 43 and raise the third finger; count aloud 44 and raise the fourth finger; count aloud 45 and raise the fifth finger; count aloud 46 and raise the sixth finger; count aloud 47 and raise the seventh finger; count 48 and raise the eighth finger; count aloud 49 and raise the ninth finger; and count aloud 50 and raise the tenth finger.

40,

Count up 1 space: 41

Count up 2 spaces: 42

Count up 3 spaces: 43

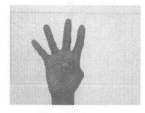
Count up 4 spaces: 44

Count up 5 spaces: 45

Count up 6 spaces: 46

Count up 7 spaces: 47

Count up 8 spaces: 48

Count up 9 spaces: 49

Count up 10 spaces: 50

Count up a total of 10 spaces
$10 \times 5 = 50$

56

In this example the table of 10 will be used. Students will begin writing out the table of 10, beginning with the multiple of 1 and continuing until the target number of 50 is reached.

$10 \times 1 = 10 \quad 10 \times 5 = 50$

$10 \times 2 = 20$

$10 \times 3 = 30$

$10 \times 4 = 40$

In the next example, the table of 12 will be used to determine the number of times 12 will divide into 48. The two-digit problems will help students understand how to apply the counting-up strategy when using two-digit numbers in an equation.

Counting Up Using the Time Table of 12

As students prepare to count up using the time table of 12, explain to them that once they have counted up 10 spaces, they will have to use one of their hands a second time. Instruct students to remember the last number that was counted, write this number on a scratch sheet of paper if this will help them remember it, close both fists, and continue counting up by raising a finger to represent the 11th space and then raising another finger a second time to represent the 12th space. Note the illustration below.

$48 \div 12 = ?$

Identify the target number students will be searching for by making a clear statement: "The target number we will be searching for in this equation is 48." Explain we are trying to determine the number of times 12 will divide into 48.

To determine the number of times 12 will divide into 48, we will list the time table of 12 until we find the number that, when multiplied with 12, produces the product 48.

Step 1: Tell students to remember the rule for multiplying by 1: any number multiplied by 1 equals that number.

$12 \times 1 = 12$

Step 2: $12 \times 2 = ?$

Begin counting aloud with the number 12. Next, count aloud 13 and raise one finger; count aloud 14 and raise the second finger; count aloud 15 and raise the third finger; count aloud 16 and raise the fourth finger; count aloud 17 and raise the fifth finger; count aloud 18 and raise the sixth finger; count aloud 19 and raise the seventh finger; count 20 and raise the eighth finger; count aloud 21 and raise the ninth finger; count aloud 22 and raise the tenth finger (at this point, students must recognize that they have used all ten fingers, so one hand will be used a second

57

time; student must remember the last number counted and make a note of it if necessary); count aloud 23 and raise one finger a second time; and count aloud 24 and raise another finger a second time.

12,

Count up 1 space: 13

Count up 2 spaces: 14

Count up 3 spaces: 15

Count up 4 spaces: 16

Count up 5 spaces: 17

Count up 6 spaces: 18

Count up 7 spaces: 19

Count up 8 spaces: 20

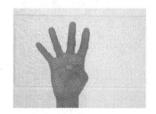

Count up 9 spaces: 21

Count up 10 spaces: 22

Count up 11 spaces: 23

Count up 12 spaces: 24

Count up a total of 12 spaces
$12 \times 2 = 24$
Step 3: Explain to students that they now know that $12 \times 2 = 24$, so now they can determine the product of 12×3.

$12 \times 3 = ?$
Begin counting aloud with the number 24; next count aloud 25 and raise one finger; count aloud 26 and raise the second finger; count aloud 27 and raise the third finger; count aloud 28 and raise

58

the fourth finger; count aloud 29 and raise the fifth finger; count aloud 30 and raise the sixth finger; count aloud 31 and raise the seventh finger; count 32 and raise the eighth finger; count aloud 33 and raise the ninth finger; count aloud 34 and raise the tenth finger (at this point, students must recognize that they have used all ten fingers, so one hand will be used a second time; remember to make a note of the last number counted); count aloud 35 and raise one finger a second time; and count aloud 36 and raise another finger a second time.

24,

Count up 1 space: 25

Count up 2 spaces: 26

Count up 3 spaces: 27

Count up 4 spaces: 28

Count up 5 spaces: 29

Count up 6 spaces: 30

Count up 7 spaces: 31

Count up 8 spaces: 32

Count up 9 spaces: 33

Count up 10 spaces: 34

Count up 11 spaces: 35

Count up 12 spaces: 36

Count up a total of 12 spaces

$12 \times 3 = 36$

Explain to students that they now know that $12 \times 3 = 36$, so now they can determine the product of 12×4.

$12 \times 4 = ?$

Begin counting aloud with the number 36. Next, count aloud 37 and raise one finger; count aloud 38 and raise the second finger; count aloud 39 and raise the third finger; count aloud 40 and raise the fourth finger; count aloud 41 and raise the fifth finger; count aloud 42 and raise the sixth finger; count aloud 43 and raise the seventh finger; count 44 and raise the eighth finger; count aloud 45 and raise the ninth finger; count aloud 46 and raise the tenth finger (at this point, students must recognize that they have used all ten fingers, so one hand will be used a second time; make a note of the last number counted); count aloud 47 and raise one finger a second time; and count aloud 48 and raise another finger a second time.
36,

Count up 1 space: 37

Count up 2 spaces: 38

Count up 3 spaces: 39

Count up 4 spaces: 40

Count up 5 spaces: 41

Count up 6 spaces: 42

Count up 7 spaces: 43

Count up 8 spaces: 44

Count up 9 spaces: 45

Count up 10 spaces: 46

Count up 11 spaces: 47

Count up 12 spaces: 48

Count up a total of 12 spaces

$12 \times 4 = 48$

Target number = 48

In this example, the table 12 is used. Students will begin writing out the table of 12, beginning with the multiple of 1 and continuing until the target number of 48 is identified.

$12 \times 1 = 12$

$12 \times 2 = 24$

$12 \times 3 = 36$

$12 \times 4 = 48$

Counting Up Using the Time Table of 11

As students prepare to count up using the time table of 11, explain to students that once they have counted up 10 spaces, they will have to use one of their hands a second time. Instruct students to remember the last number that was counted, making a note of this number on scratch paper if necessary. Instruct students to close both fists, and then continue counting up by raising one finger a second time to represent the 11th space.

$66 \div 11 = ?$

Identify the target number students will be searching for by making a clear statement: "The target number that we will be searching for in this equation is 66." Explain that we are trying to determine the number of times 11 will divide into 66.

To determine the number of times 11 will divide into 66, we will list the time table of 11, and continuing until we find the number that, when multiplied with 11, will produce a product of 66.

Step 1: Tell students to remember the rule for multiplying by 1: any number multiplied by 1 equals that number.

$11 \times 1 = 11$

Step 2: Explain to students that this is where they will begin to count up to determine the product of $11 \times 2 = ?$

Begin counting aloud with the number 11. Next, count aloud 12 and raise one finger; count aloud 13, and raise the second finger; count aloud 14 and raise the third finger; count aloud 15 and raise the fourth finger; count aloud 16 and raise the fifth finger; count aloud 17 and raise the sixth finger; count aloud 18 and raise the seventh finger; count aloud 19 and raise the eighth finger; count aloud 20 and raise the ninth finger; count aloud 21 and raise the tenth finger (at this point, students must recognize that they have used all ten fingers, so one hand will be used a second time; make a note of the last number counted, and then continue counting); and count aloud 22 and raise one finger a second time.

11,

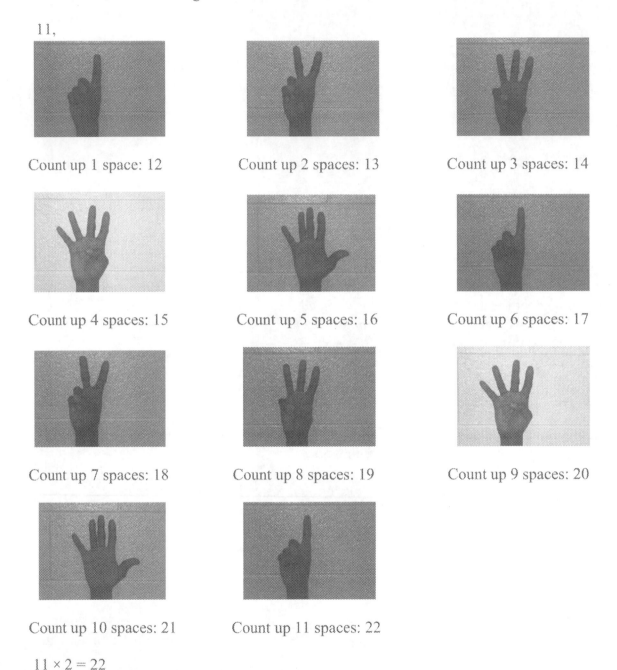

Count up 1 space: 12   Count up 2 spaces: 13   Count up 3 spaces: 14

Count up 4 spaces: 15   Count up 5 spaces: 16   Count up 6 spaces: 17

Count up 7 spaces: 18   Count up 8 spaces: 19   Count up 9 spaces: 20

Count up 10 spaces: 21   Count up 11 spaces: 22

$11 \times 2 = 22$

Step 3: Explain to students that they now know that $11 \times 2 = 22$, so now they can determine the product of 11×3.

$11 \times 3 = ?$

Begin counting aloud with the number 22. Next, count aloud 23 and raise one finger; count aloud 24 and raise the second finger; count aloud 25 and raise the third finger; count aloud 26 and raise the fourth finger; count aloud 27 and raise the fifth finger; count aloud 28 and raise the sixth finger; count aloud 29 and raise the seventh finger; count 30 and raise the eighth finger; count aloud 31 and raise the ninth finger; count aloud 32 and raise the tenth finger (at this, point students must recognize that they have used all ten fingers, so one hand will be used a second time; make a note of the last number counted, and continue counting); and count aloud 33 and raise one finger a second time.

22,

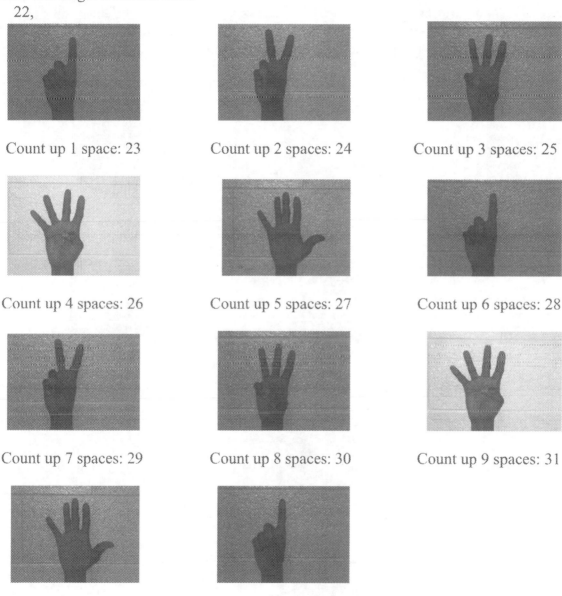

Count up 1 space: 23   Count up 2 spaces: 24   Count up 3 spaces: 25

Count up 4 spaces: 26   Count up 5 spaces: 27   Count up 6 spaces: 28

Count up 7 spaces: 29   Count up 8 spaces: 30   Count up 9 spaces: 31

Count up 10 spaces: 32   Count up 11 spaces: 33

Count up a total of 11 spaces
11 × 3 = 33

Step 4: Explain to students that they now know that 11 × 3 = 33, so now they can determine the product of 11 × 4.

63

$11 \times 4 = ?$

Begin counting aloud with the number 33. Next, count aloud 34 and raise one finger; count aloud 35 and raise the second finger; count aloud 36 and raise the third finger; count aloud 37 and raise the fourth finger; count aloud 38 and raise the fifth finger; count aloud 39 and raise the sixth finger; count aloud 40 and raise the seventh finger; count 41 and raise the eighth finger; count aloud 42 and raise the ninth finger; count aloud 43 and raise the tenth finger (at this point, students must recognize that they have used all ten fingers, so one hand will be used a second time; make a note of the last number counted, and continue counting); and count aloud 44 and raise one finger a second time.

33,

Count up 1 space: 34

Count up 2 spaces: 35

Count up 3 spaces: 36

Count up 4 spaces: 37

Count up 5 spaces: 38

Count up 6 spaces: 39

Count up 7 spaces: 40

Count up 8 spaces: 41

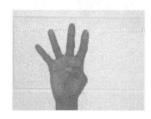

Count up 9 spaces: 42

Count up 10 spaces: 43

Count up 11 spaces: 44

Count up a total of 11 spaces
$11 \times 4 = 44$

Step 5: Explain to students that they now know that $11 \times 4 = 44$, so now they can determine the product of 11×5.

$11 \times 5 =?$

Begin counting aloud with the number 44. Next, count aloud 45 and raise one finger; count aloud 46 and raise the second finger; count aloud 47 and raise the third finger; count aloud 48 and raise the fourth finger; count aloud 49 and raise the fifth finger; count aloud 50 and raise the sixth finger; count aloud 51 and raise the seventh finger; count 52 and raise the eighth finger; count aloud 53 and raise the ninth finger; count aloud 54 and raise the tenth finger (at this point, students must recognize that they have used all ten fingers, so one hand will be used a second time; make a note of the last number counted, and continue counting); and count aloud 55 and raise one finger a second time.

44,

Count up 1 space: 45

Count up 2 spaces: 46

Count up 3 spaces: 47

Count up 4 spaces: 48

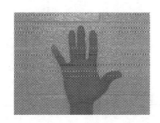

Count up 5 spaces: 49

Count up 6 spaces: 50

Count up 7 spaces: 51

Count up 8 spaces: 52

Count up 9 spaces: 53

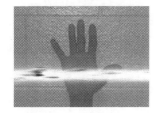

Count up 10 spaces: 54

Count up 11 spaces: 55

Count up a total of 11 spaces

$11 \times 5 = 55$

Step 6: Explain to students that they now know that $11 \times 5 = 55$, so now they can determine the product of 11×6.

$11 \times 6 = ?$

Begin counting aloud with the number 55. Next, count aloud 56 and raise one finger; count aloud 57 and raise the second finger; count aloud 58 and raise the third finger; count aloud 59 and raise the fourth finger; count aloud 60 and raise the fifth finger; count aloud 61 and raise the sixth finger; count aloud 62 and raise the seventh finger; count 63 and raise the eighth finger; count aloud 64 and raise the ninth finger; count aloud 65 and raise the tenth finger (at this point, students must recognize that they have used all ten fingers, so one hand will be used a second time; make a note of the last number counted, and continue counting); and count aloud 66 and raise one finger a second time.

55,

Count up 1 space: 56

Count up 2 spaces: 57

Count up 3 spaces: 58

Count up 4 spaces: 59

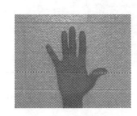
Count up 5 spaces: 60

Count up 6 spaces: 61

Count up 7 spaces: 62

Count up 8 spaces: 63

Count up 9 spaces: 64

Count up 10 spaces: 65   Count up 11 spaces: 66

Count up a total of 11: spaces
$11 \times 6 = 66$

In this example, the table of 11 is used. Students will begin writing out the table of 11, beginning with the number 1 and continuing until the target number of 66 is reached.

$11 \times 1 = 11$ $11 \times 5 = 55$

$11 \times 2 = 22$ $11 \times 6 = 66$

$11 \times 3 = 33$

$11 \times 4 = 44$

It is very important to have students manually write out the time table chart each time they begin to solve a multiplication or division equation. Students should always begin to solve a problem using the first multiple of 1 and continuing until the target number or product is reached. The repetition and practice of manually writing out the time tables will help students learn multiplication, division, and time tables simultaneously.

Scratch Paper:

Chapter 6
Counting Up Strategy

Chapter 7 is intended to be a step-by-step instructional guide to using the counting-up strategy to solve multiplication, division, and subtraction equations. The strategy will also help students learn the time table chart in its entirety.

Step 1: Identify the problem (s) to be solved. Identify the target number students will need to identify to solve the equation.

$12 \div 6 = ?$

Note the division problem we will be solving. The target number we are looking for is 12. The target number of 12 is the product of 6 multiplied by a second number; students will determine the missing multiple.

Step 2: Identify the time table multiple to be used to solve the division problem (this number is always the divisor).

Step 3: In this equation, the multiple of 6 is used; we will be counting up 6 spaces from the product of 6×1 (which is 6).

Step 4: We begin by saying $6 \times 1 = 6$, reminding students of the rule of multiplying by 1, (one multiplied by any number is that number).

Count aloud 6.

Step 5: Count aloud 7 and raise one finger.

Step 6: Count aloud 8 and raise the second finger.

Step 7: Count aloud 9 and raise the third finger.

Step 8: Count aloud 10 and raise the fourth finger.

Step 9: Count aloud 11 and raise the fifth finger.

Step 10: Count aloud 12 and raise the sixth finger.

We have counted up 6 spaces from the product of 6 to 12, so $12 \div 6 = 2$, because $6 \times 2 = 12$.

In this step, students will use the table of 6. Students will manually write out the table of 6, beginning with the first multiple of 1 and continuing until the target number of 12 is reached.

$6 \times 1 = 6$

$6 \times 2 = 12$

When the product of 12 is identified, the equation is complete; the inverse operation of $12 \div 6$ is used to help solve the division problem. Students use multiplication to determine which factor, when multiplied by 6, is equivalent to 12. $6 \times 2 = 12$, so $12 \div 6 = 2$.

Chapter 7
Counting-Up Strategy and Subtraction

One of the most appealing things about the counting-up strategy is that it can be applied to various areas of math computation. The examples that follow will explain how the strategy can be applied to subtraction problems. In the previous examples and illustrations, we discussed how to add or count numbers in a numerical sequence that always *increased* in value; now the numerical value of the numbers used in the sequences will always *decrease* in value.

$42 - 6 =$ $72 - 8 =$ $49 - 7 =$ $4 - 2 =$

$3 - 3 =$ $50 - 10 =$ $48 - 12 =$ $66 - 11 =$

We will begin with the equation

$42 - 6 = ?$

Students will begin with the number 42 and then count down 6 spaces in order to subtract 6 from 42. Begin by saying the number 42 aloud; next, say, "Now let's count 6 spaces backward away from 42."

Count aloud 42, now count down 6 spaces.

Count down 1 space: 41

Count down 1 space: 40

Count down 1 space: 39

Count down 1 space: 38

Count down 1 space: 37

Count down 1 space: 36

We have counted down to the sum of 36, so if we subtract 6 from 42, the sum is 36:

$42 - 6 = 36$

Explain to students that every time we count down a space, one finger is folded down to show that 1 has been subtracted; until a total of 6 spaces have been counted down to represent the number that is being subtracted.

2. Now let's subtract:

72 – 8 =?

Let's identify the number of spaces that we will be counting down by making a clear statement: "The number of spaces that we will be counting down in this problem is 8."

We will begin with the equation

72 – 8 =?

Students will begin with the number 72 and then count down 8 spaces in order to subtract 8 from 72. Begin by saying the number 72 aloud; next, say, "Now, let's count 8 spaces backward down from 72." Tell students be sure to fold down one finger each time a space is counted down.

Count aloud 72; now count down 8 spaces.

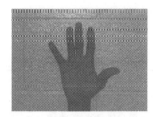

Count down 1 space: 71

Count down 1 space: 70

Count down 1 space: 69

Count down 1 space: 68

Count down 1 space: 67

Count down 1 space: 66

Count down 1 space: 65   Count down 1 space: 64

We have counted down from 72 to the sum of 64, so if we subtract 8 from 72, the sum is 64:

$72 - 8 = 64$

Explain to students that every time we count down one space, one finger is folded down to show that 1 has been subtracted, until a total of 8 spaces have been counted down to represent the number that is being subtracted.

Now let's subtract:

$49 - 7 = ?$

Let's identify the number of spaces that we will be counting down by making a clear statement: "The number of spaces that we will be counting down in this problem is 7."

We will begin with the equation

$49 - 7 = ?$

Students will begin with the number 49 and then count down 7 spaces in order to subtract 7 from 49. Begin by saying the number 49 aloud; next, say "Now let's count 7 spaces backward down from 49." Tell students be sure to fold down one finger each time a space is counted down.

Count down 1 space: 48   Count down 1 space: 47   Count down 1 space: 46

72

Count down 1 space: 45

Count down 1 space: 44

Count down 1 space: 43

Count down 1 space: 42

We have counted down from 49 to the sum of 42, so if we subtract 7 from 49, the sum is 42:

$49 - 7 = 42$

Explain to students that every time we count down 1 space, one finger is folded down to show that 1 has been subtracted, until a total of 7 spaces have been counted down to represent the number that is being subtracted.

We will begin with the equation

$4 - 2 =?$

Identify the number of spaces that students will be counting down by making a clear statement: "The number of spaces that we will be counting down in this problem is 2."

Students will begin with the number 4 and then count down 2 spaces in order to subtract 2 from 4. Begin by saying the number 4 aloud; next, say, "Now let's count 1 space backward down from 4. Repeat this step until you have counted down a total of 2 spaces."

Count down 1 space: 4

Count down 1 space: 3

Count down 1 space: 2

Students will begin with the number 4 and then count down 2 spaces in order to subtract 2 from 4. Begin by saying the number 4 aloud; next, say, "Now let's count 1 space backward down from 4. Repeat this step until you have counted down a total of 2 spaces."

We have counted down from 4 to the sum of 2, so if we subtract 2 from 4, the sum is 2:

$4 - 2 = 2$

Explain to students that every time we count down 1 space, one finger is folded down to show that 1 has been subtracted, until a total of 2 spaces have been counted down to represent the number that is being subtracted.

4. Now let's subtract:

$3 - 3 = ?$
We will begin with the equation

$3 - 3 = ?$

Identify the number of spaces that students will be counting down by making a clear statement: "The number of spaces that we will be counting down in this problem is 3."

Students will begin with the number 3 and then count down 3 spaces in order to subtract 3 from 3. Begin by saying the number 3 aloud; next, say "Now let's count 1 space backward down from 3. Repeat this step until you have counted down a total of 3 spaces."

Count down 1 space: 3

Count down 1 space: 2

Count down 1 space: 1

Count down 1 space: 0

We have counted down from 3 to the sum of 0, so if we subtract 3 from 3, the sum is 0:

$3 - 3 = 0$

Explain to students that every time we count down 1 space, one finger is folded down to show that 1 has been subtracted, until a total of 3 spaces have been counted down to represent the number that is being subtracted.

Now let's subtract:

$50 - 10 = ?$

Identify the number of spaces that students will be counting down by making a clear statement: "The number of spaces that we will be counting down in this problem is 10."

We will begin with the equation
$50 - 10 = ?$

Students will begin with the number 50 and then count down 10 spaces in order to subtract 10 from 50. Begin by saying the number 50 aloud; next, say, "Now let's count 1 space backward down from 50. Repeat this step until you have counted down a total of 10 spaces."

Count down 1 space: 49

Count down 1 space: 48

Count down 1 space: 47

Count down 1 space: 46

Count down 1 space: 45

Count down 1 space: 44

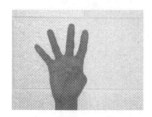
Count down 1 space: 43

Count down 1 space: 42

Count down 1 space: 41

Count down 1 space: 40

Now we have counted down 10 spaces to 40.

We have counted down from 50 to the sum of 40, so if we subtract 10 from 50, the sum is 40:

50 – 10 = 40

Explain to students that every time we count down 1 space, one finger is folded down to show that 1 has been subtracted, until a total of 10 spaces have been counted down to represent the number that is being subtracted.

Now let's subtract:

48 – 12 =?

Identify the number of spaces that students will be counting down by making a clear statement: "The number of spaces that we will be counting down in this problem is 12."

Students will begin with the number 48 and then count down 12 spaces in order to subtract 12 from 48. Begin by saying the number 48 aloud; next, say "Now let's count 1 space backward down from 48. Repeat this step until you have counted down a total of 12 spaces."

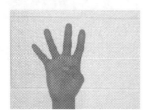
Count down 1 space: 47

Count down 1 space: 46

Count down 1 space: 45

Count down 1 space: 44

Count down 1 space: 43

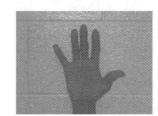
Count down 1 space: 42

Count down 1 space: 41

Count down 1 space: 40

Count down 1 space: 39

Count down 1 space: 38   Count down 1 space: 37   Count down 1 space: 36

Now we have counted down a total of 12 spaces to the sum of 36:

48 – 12 = 36

Explain to students that every time we count down 1 space, one finger is folded down to show that 1 has been subtracted until a total of 12 spaces have been counted down to represent the number that is being subtracted.

Now let's subtract:

66 – 11 =?

Identify the number of spaces that students will be counting down by making a clear statement: "The number of spaces that we will be counting down in this problem is 11."

Students will begin with the number 66 and then count down 11 spaces in order to subtract 11 from 66. Begin by saying the number 66 aloud; next, say "Now let's count 1 space backward down from 66. Repeat this step until you have counted down a total of 11 spaces."
66,

Count down 1 space: 65   Count down 1 space: 64   Count down 1 space: 63

Count down 1 space: 62   Count down 1 space: 61   Count down 1 space: 60

Count down 1 space: 59

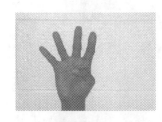

Count down 1 space: 58

Count down 1 space: 57

Count down 1 space: 56

Count down 1 space: 55

Now we have counted down a total of 11 spaces to the sum of 55:

$66 - 11 = 55$

Explain to students that every time we count down 1 space, one finger is folded down to show that 1 has been subtracted, until a total of 11 spaces have been counted down to represent the number that is being subtracted.

Conclusion
What Can You Expect Your Students to Learn from This Strategy?

This strategy is a reliable and effective method of teaching students basic math skills that they can rely on for years to come. If the strategy is introduced, followed consistently, and practiced on a routine basis, students will be able to perform basic math computation on a proficient level.

1. Students will learn basic math skills:

a. Multiplication
b. Division
c. Time table chart

2. Students will learn three basic math skills simultaneously.

3. Students will learn how to use their fingers as counting tools on a temporary basis.

4. Students will learn multiplication factors from 1 to 12.

5. Students will learn the rule of multiplying using 0 as a factor.

6. Students will learn the rule of multiplying using 1 as a factor.

7. Students will learn the rule of dividing by 0.

8. Students will learn the rule of dividing a number by itself.

9. Students will learn to solve a basic long-division equation.

10. Students will learn to use visual, auditory, and trial-and-error approaches to learn.

11. Students will learn the concept of inverse operation.

12. Students will learn to apply the strategy to subtraction equations.

Once students learn the strategy of counting up, they will be able to apply the strategy to multiplication, division, subtraction, and the time table chart. Students can be expected to develop to a level of proficient performance where the use of their fingers for counting will no longer be necessary.

Instruction Practice Page

Follow the steps listed below to fill in the missing information.

$24 \div 4 = ?$

In this equation, the goal is to determine the number of times 4 will divide into 24.

Step 1: Determine which number is the target number. Remember, the target number is the number students will need to identify.

The target number in this equation is _____.

Step 2: Determine which multiplication table will be used to solve the equation.

In this equation, the table of _____ will be used, because it must be determined how many times _____ will divide into _____.

Step 3: Begin to manually write the table of _____, beginning with the first multiple of the

table of 4, which will be _____. This step continues until the second multiple is identified

that will give a product of _____ when multiplied by 4.

$4 \times$ _____ $= 24$

Step 4: Apply the strategy of counting up using your fingers to help determine the next product in the time table, which is $4 \times$ _____.

$4 \times$ _____ $=$ _____

Begin with the number _____ and count up _____ spaces, raising one finger as each number is counted aloud.

Begin counting with the number 4 and count up _____ spaces.

_____   _____   _____   _____   _____

At this point, you have counted up from 4 (the product 4×1) to determine that $4 \times 2 =$ _____. Manually list the new product.

_____ $\times$ _____ $=$ _____

_____ × _____ = _____

Now repeat the process to determine the product of the next factor, which is _____. To determine the new product, count up 4 spaces from the product of 8. Students will raise _____ finger as they count up each space.

_____ × 3 = ?

Begin counting with the product of _____ and count up 4 spaces.

8, _____ _____ _____ _____

At this point, you have counted up from 8 (the product of 4 × 2) to determine that 4 × 3 = _____. Manually list the new product.

_____ × _____ = _____

_____ × _____ = _____

_____ × _____ = _____

Begin counting with the product of 12 and count up _____ spaces.

_____ _____ _____ _____ _____

At this point, you have counted up 4 spaces from 12 (the product of 4 × 3) to determine that 4 × 4 = _____. Manually list the new product.

_____ × _____ = _____   _____ × _____ = _____

_____ × _____ = _____

_____ × _____ = _____

Now repeat the process to determine the product of the next factor, which is _____. To determine the new product, count up 4 spaces from the product of 16. Students will raise _____ finger as they count up each space.

_____ × _____ = ?

Begin counting with the number _____ and count up 4 spaces.

16, _____ _____ _____ _____

At this point, you have counted up from 16 (the product 4 × 4) to determine that 4 × 5 = _____. Manually list the new product.

_____ × _____ = _____   _____ × _____ = _____

_____ × _____ = _____   _____ × _____ = _____

_____ × _____ = _____

Now repeat the process to determine the product of the next factor, which is _____. To determine the new product, count up 4 spaces from the product of 20. Students will raise _____ finger as they count up each space.

_____ × _____ = ?

Begin counting with the number _____ and count up 4 spaces.

20, _____   _____   _____   _____

At this point, you have counted up 20 (the product 4 × 5) to determine that 4 × 6 = _____. Manually list the new product.

_____ × _____ = _____   _____ × _____ = _____

_____ × _____ = _____   _____ × _____ = _____

_____ × _____ = _____   _____ × _____ = _____

The target number of _____ has been identified. The product of 4 × 6 = _____, so use the inverse operation of 4 × 6 = _____ to determine the missing factor.

24 ÷ 4 = _____.

Close examination of this practice page reveals the answers to each equation.

Scratch Paper:

About the Author

Andray McCuien was born in Little Rock, Arkansas, and now resides in Detroit, Michigan. He graduated from Hall High School in 1977 and attended the University of Arkansas in Little Rock. Majoring in art was his first choice but not his passion. He later completed his bachelor of art degree in June 1994, at the University of Central Arkansas. In 1998 he moved to Michigan; he began teaching third-graders at Hosmer Elementary School in the Detroit public-school system in 2000.

After three years, he began teaching at Noble Middle School, where he taught language arts and social studies to sixth-, seventh-, and eighth-graders. The next year, he had the privilege of teaching children with learning disabilities. He decided to pursue his master's degree in special education with an institutional recommendation. He received an award in June 2008 from Grand Canyon University, located in Phoenix, Arizona.

Mr. McCuien noticed the students were not learning their math skills at a normal pace. This frustrated him as a teacher, so he developed a new, nontraditional way of teaching, learning, and remembering math.

After he coined the concept of counting up, his students began to learn and understand the multiplication system much faster. The students were able to learn division and solve math problems just by counting up. Andray collected the time table charts and began working with his students by making them count on their hands. The strategy of counting up was born.

Reader Notes

Reader Notes